Exploration Into
THE POLAR REGIONS

DAVID ROOTES

Chelsea House Publishers
Philadelphia

First published in hardback edition in 2001
by Chelsea House Publishers, a subsidiary of
Haights Cross Communications. All rights reserved.
Printed and bound in China.

First published in the UK in 1994 by
Belitha Press Limited, London House,
Great Eastern Wharf, Parkgate Road,
London SW11 4NQ, England

Editor: Jill Laidlaw
Designer: Simon Borrough
Picture Researcher: Juliet Duff
Series consultant: Shane Winser, Royal Geographical Society,
London
Consultants: Ann Shirley formerly of the National Maritime
Museum
Map annotations: Hardlines

First printing
1 3 5 7 9 8 6 4 2

The Chelsea House World Wide Web address is
http://www.chelseahouse.com

Library of Congress Cataloging-in-Publication Data applied for.

ISBN: 0-7910-6026-8

Picture acknowledgments:
Bryan and Cherry Alexander: front cover foreground, 1,
6 bottom, 9 top, 29 bottom, 41 bottom, 45 right. Ancient
Art and Architecture Collection: 28 top. Bridgeman Art
Library: 2, 7 top british Library, 18 center Giraudon,
19 bottom, 30 bottom. British Antarctic Survey: 39
bottom, 43 bottom. E.T. Archive: 14 right, 20 center,
21 bottom, 24 top. Mary Evans Picture Library: 7
bottom, 25 bottom, 27 center. Werner Forman Archive:
6 top, 15 bottom, 17 top. John Hancock Mutual Life
Insurance Company, Boston, Massachusetts: 26 bottom.
Robert Harding Picture Library: front cover background,
13 center, 23 top left, 39 top. Michael Holford: 29 top.
Hulton Deutsch Collection: 21 top, 37 bottom and inset.
Mansell Collection: 14 left, 22 right, 35 bottom. NASA:
45 left, Nasjonalgalleriet, Oslo: 16 bottom. National
Maritime Museum, Greenwich: 23 bottom, 30 top, 32
top. Peter Newark's Pictures: 17 bottom, 19 top, 24
center, 28 bottom. Novosti Photo Library: 31 bottom,
41 top. Oxford Scientific Films Ltd: 11 top left ben
Osborne, 13 top Doug Allan, 33 bottom Kim Westerskov,
43 top Colin Monteath. Planet Earth Pictures: 9 bottom.
Royal Geographical Society: 11 bottom, 25 top, 26 top,
34 center, 36 left, 38 bottom, 44, back cover. Science
Photo Library: 11 top right, 42. Scott Polar Research
Institute, Cambridge: 23 top right, 27 tope left, 32
bottom, 37 center.

In 1902 Robert Scott, Ernest Shackleton and Edward Wilson set out to sledge to the South Pole. They had to turn back weakened by scurvy. Wilson painted this picture in 1903.

Contents

An Inuk from Alaska looks out over the Arctic Ocean for seals.

1 Introduction

At the Ends of the Earth

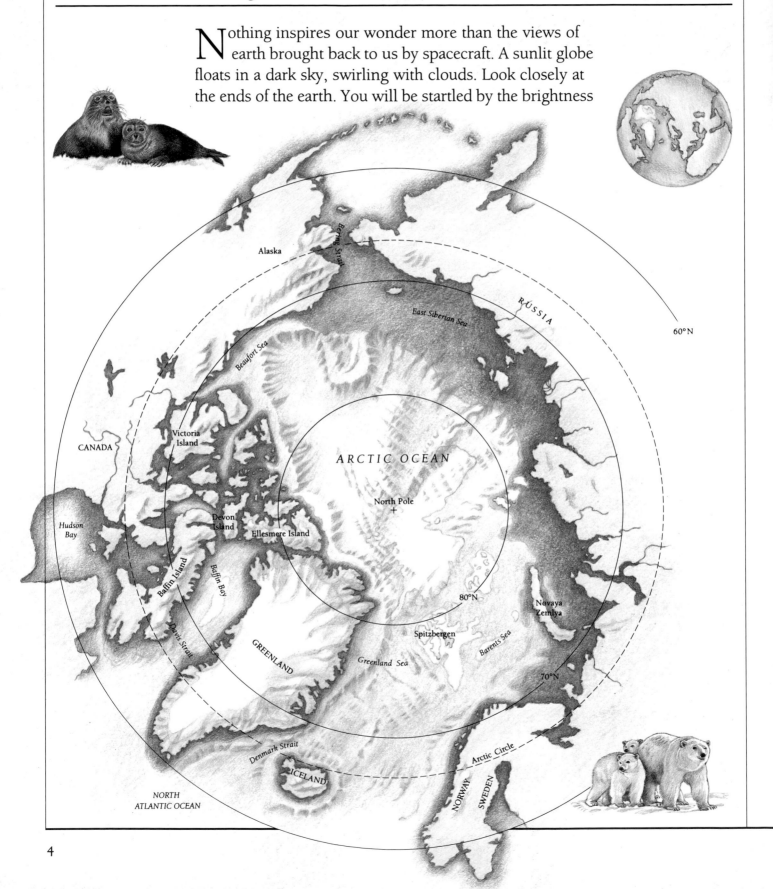

Nothing inspires our wonder more than the views of earth brought back to us by spacecraft. A sunlit globe floats in a dark sky, swirling with clouds. Look closely at the ends of the earth. You will be startled by the brightness

Alaska

Bering Strait

East Siberian Sea

RUSSIA

60°N

Beaufort Sea

Victoria
Island

CANADA

ARCTIC OCEAN

North Pole
+

Hudson
Bay

Devon
Island

Ellesmere Island

80°N

Novaya
Zemlya

Baffin Island

Baffin Bay

Spitzbergen

Barents Sea

70°N

Davis Strait

GREENLAND

Greenland Sea

Denmark Strait

Arctic Circle

ICELAND

NORWAY

SWEDEN

NORTH
ATLANTIC OCEAN

of the reflected light, because here lie deserts of ice that never melt. In the north is the frozen sea called the Arctic and in the south is the continent, Antarctica.

Exploring This Book

This book is divided into five chapters. The first one describes the differences between the Arctic and the Antarctic. The second looks at the people who live in the Arctic (no one has ever settled in the Antarctic). The next two chapters separate the exploration of the Arctic and the Antarctic. Finally we look at the polar regions today, at both their political and environmental importance.

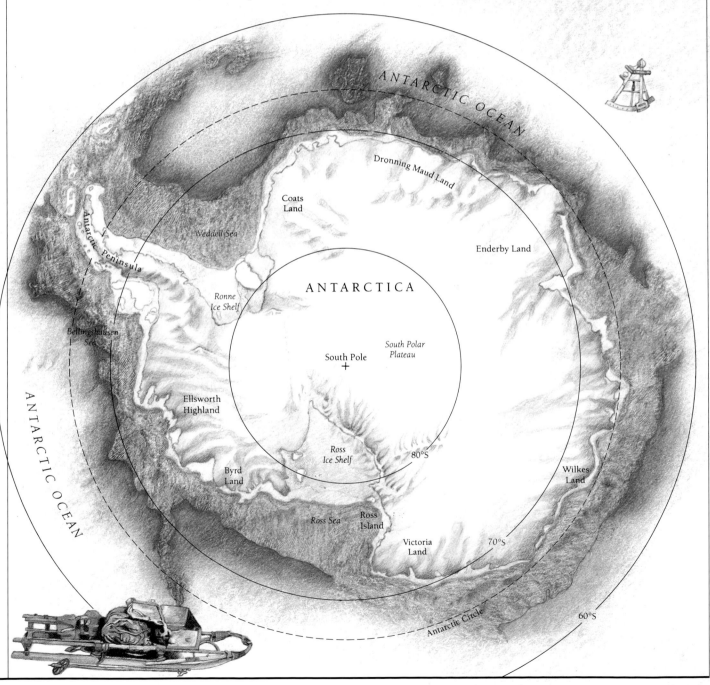

The Beginnings of Polar Discovery

This inch and a half-high mask was made 2,500 years ago by the first true Arctic people. They were called the Dorset Culture.

Inuit can travel in summer by day or night, as the sun never sets.

The poles are the coldest places on earth. Their brilliant whiteness reflects the light of the sun, so they never get warm. The lowest temperature ever recorded is −128° Fahrenheit in Antarctica. At that temperature a cup of boiling water thrown into the air freezes instantly. The poles are also very windy. The cold wind and endless expanse of ice make the **polar regions** very hostile places in which to live.

Arctic People

Because of the **climate** of the polar regions and the lack of food, it was thought that no one could live there. In fact, about 20,000 years ago people migrated north from North America and Eurasia into the Arctic. They would go north in the warmer summer months to hunt and fish, and in the winter head back south again.

The Midnight Sun

In summer at the North Pole the sun does not set for six months. It moves around the sky and is never out of sight. During the same six months, the sun never rises in the South Pole. It is in permanent darkness. For the other six months of the year the situation is reversed.

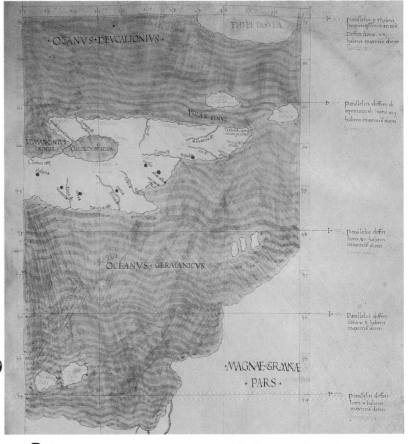

Pytheas set out over 2,000 years ago to find new lands to the north. This early map shows the land he reached, which was probably Iceland.

Early Explorers

Over 2,000 years ago Greek scholars described the world as having a freezing north, which they called Arktos. They believed that if there was a cold north, there must be a cold south. This they named the Antarktos ("ante" in Greek means opposite). A Greek trader, Pytheas, sailed north and was said to have reached Iceland, but the sea farther north was all ice. He was looking for new lands with which to trade. This was a major reason for the start of exploration.

Pythagoras was a Greek scholar who believed that the earth was round. He thought the North Pole was a frozen land.

The Arctic

Stand at the North Pole and you stand on a frozen sea over a deep ocean. There is no land in sight. In fact, the nearest land is 400 miles away in north Greenland. You are standing at the center of a large and almost completely enclosed ocean and you are exactly 90° North (see the box on the right for an explanation of this measurement). Whichever direction you look is south and within a few minutes you can walk around the world.

The Arctic has several boundaries. It includes a large ocean that is almost enclosed by North America, Russia, and Europe.

Huskies have been used by Inuit for hundreds of years for traveling in the Arctic. Not only are they good friends, but they will also pull a heavy sled for many miles.

Beneath your feet is ice which is always moving. If you stand still, you will gradually be swept along by the **ocean currents**. You may also get very wet because **sea ice** is unpredictable. Many explorers have been terrified by seeing the ice breaking up around their camp, exposing them to the frigid Arctic Ocean.

Arctic Boundaries

Scientists argue about what is the best Arctic boundary. Some use the Arctic Circle. A more useful boundary is the northern limit of the tree line. Beyond this line it is too cold for trees to grow.

The tree line is an obvious frontier and an important one for animals. Beyond the tree line is the **tundra**, where there is no shelter in winter.

But the tree line means nothing at sea. Here an important boundary is made by the sea ice. In winter the ice spreads far into the Atlantic and Bering seas. A line can be drawn on maps showing the farthest south that ice will spread in winter.

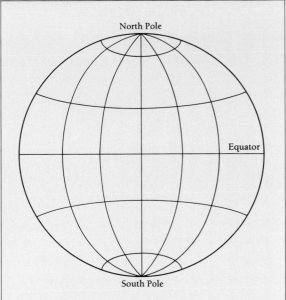

Lines of latitude and longitude on the globe

Icebergs are bits of glaciers that have broken off into the sea. An iceberg has more than twice as much ice under the water.

Under the Ice

The great land masses of the northern **hemisphere** form the edges of the Arctic Ocean, with North America to one side and Eurasia on the other. The gaps are nearly filled by islands such as Greenland and Spitzbergen. The North Atlantic and Bering Strait are the only way in by ship. Early explorers in small wooden vessels braved these passages. More recently a Russian nuclear-powered icebreaker was able to push its way across the Arctic Ocean, smashing a track through the sea ice. Submarines have traveled beneath the ice, surfacing near the North Pole through cracks in the ice.

Lines on the Earth

The globe can be divided up by lines encircling the earth starting at the equator, which is exactly around the middle, and going north and south. These are called lines of latitude. And the globe can be divided by lines running from the North Pole to the South Pole. These are lines of longitude. The point where lines of longitude and latitude cross gives an exact description of a position on the globe.

Latitude was known to early Greek and Arab navigators. They made simple devices to measure the angle of the sun above the horizon at midday. A calculation showed how many degrees they were away from the equator. The equator is 0° and if you travel north your latitude gradually increases to 90° at the North Pole. For each degree you travel about 70 miles.

Confusingly, the same thing happens when you travel south from the equator. Latitude must always have north (N) or south (S) added to show which side of the equator you are. So, London is 51°N, and the North Pole 90°N. But Sydney, Australia, is 33°S, and the South Pole 90°S.

To work out the exact position, degrees are broken down into **minutes**. Just as there are 60 minutes in an hour, so there are 60 minutes in a degree.

The Arctic Ocean is covered by sea ice most of the year. This ice is always moving and can suddenly crack open or close up.

The Antarctic

The South Pole is one of the coldest places on earth. It is much colder than the North Pole. Even in summer the temperature never rises above freezing. Usually it is −4°F or −22°F. In winter it can be twice as cold as that. It is so cold that there is hardly any snowfall. When you breathe at those temperatures, your teeth hurt. You must always wear a mask over your face, but the water in your breath soon ices this up.

What is under your feet is stranger still. It looks like the snow and ice at the North Pole. But here the ice is more than one and a half miles thick. No wonder it is so cold.

How Big Is Antarctica?

Antarctica is an immense land. It is twice the size of Australia. The United States and Mexico would easily fit inside it. When you stand at the South Pole you are nearly 1,240 miles from the sea.

Scientists calculate that Antarctica contains nearly three-quarters of all the fresh water in the world. All of this water is frozen into the gigantic **ice cap** that covers Antarctica. Whole mountain chains have been covered by the ice cap. Only the tops of the peaks show through. These bare rocks are called nunataks.

The small areas of ice-free rock around the coast of Antarctica are most precious. Here millions of penguins and seabirds breed. These birds need bare rock on which to nest. Thousands of birds squeeze up together on the few areas of rock.

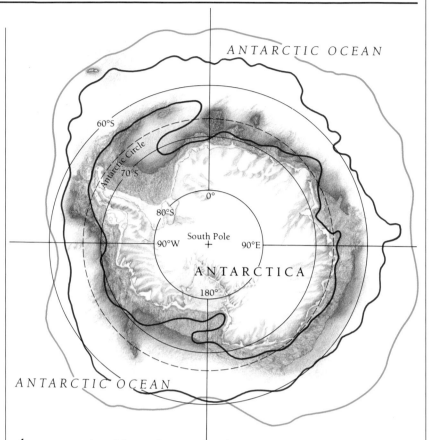

ANTARCTIC OCEAN
ANTARCTIC OCEAN
60°S
Antarctic Circle
70°S
80°S
0°
South Pole
90°W 90°E
180°
ANTARCTICA

Antarctica is the coldest and most isolated continent on earth. It is also very high—on average between 6,500 and 10,000 feet above sea level. Freezing winds howl across the landscape with few natural barriers to prevent them.

— Antarctic Polar Front
— Sea ice in summer
— Sea ice in winter

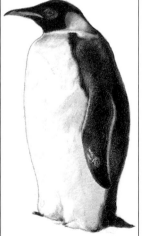

The king penguin is more than three feet tall. It keeps its egg or chick warm on its feet, not on a nest.

Antarctic Boundaries

The boundaries of the Antarctic are much easier to define than those of the Arctic. The Antarctic Ocean separates Antarctica from all the other continents. South America is the closest but is still 465 miles away. Africa is 2,500 miles from the coast of Antarctica.

The ice cap and sea ice make the Antarctic Ocean very cold. There is a point where the cold water from Antarctica meets the warm water from the Pacific, Atlantic, and Indian oceans. This is called the Antarctic Polar Front. The Polar Front runs around the Antarctic and marks a clear edge to the continent.

The Antarctic ice cap creeps slowly into the sea. Great pieces break off and drift away as tabular icebergs. The largest can be sixty miles long.

Only a few thousand scientists and helpers work in Antarctica each year. They live in comfortable buildings like the Amundsen-Scott station at the South Pole. Here the ice is 10,000 feet thick. Scientists have drilled into it to discover Antarctica's history. They found that 250 million years ago it was ice free. Then, dinosaurs roamed where now there is only ice.

Cooking equipment used in Antarctica

Antarctic exploration is a cold business. Thick, warm clothing must be worn. Even then your breath freezes to your face. The bright sunlight can damage your eyes. Any metal you touch will freeze to your skin. In winter the weather is even worse, and nobody travels.

No Home for People

The Antarctic Ocean is so wide that ancient people never reached Antarctica. If they had they would have found a land so cold that they could not have stayed.

Many early explorers looking for Antarctica were lost among the ice and waves. Eventually the continent was found. Only in the last hundred years have explorers found a way of living through the winter in Antarctica. Even now people don't live their whole lives there. Only scientists and their helpers stay in Antarctica and they only come for a year or two at the most.

2 Arctic People

Early Settlers

Some of the earliest humans, called **Neanderthals**, lived on the edges of the Arctic. They made stone tools, hunted large animals called mammoths, and used fires to keep themselves warm. About 20,000 years ago more advanced people moved up from Eurasia and North America. Neither group was able to live all year round in the Arctic.

North America has a large tundra, or treeless area, stretching from Alaska across Canada to Greenland. To live in the tundra, people had to develop skills to survive the winter as well as the summer. Eurasia has only a narrow strip of tundra. In winter people retreated into the forests for shelter. There was no reason for an Arctic culture to develop there.

The Dorset Culture

The first settled Arctic people evolved in the Alaskan tundra about 5,000 to 2,500 years ago. They were the Dorset Culture and they used small stone tools less than one inch long. Their remains have been found by archaeologists from Alaska to Greenland.

Nobody knows how these people reached North America. They may have crossed the Bering Strait from Eurasia during one of the Ice Ages. The Bering Strait is very shallow. During an Ice Age all the water would have been frozen into ice caps. This would have left the strait firm enough to walk over.

Seals provided food and oil for lamps. Clothing was made from their skins.

Musk ox and the now extinct mammoths were hunted by the earliest people in the Arctic.

The Dorset and Thule Cultures started in Alaska. They were the first people to learn how to survive the Arctic winter. They are the ancestors of the modern Inuit.

The Thule Culture

These early groups hunted musk ox, which roamed the slopes and plains of the tundra. They did not depend on the sea for their livelihood. They were gradually replaced by another culture that once again developed around the Bering Strait, particularly at St. Lawrence Island. This group hunted whales and seals as well as land animals. They had large boats covered in walrus or seal skin and could hunt in the sea ice as well as from the shore. They also made sleds for traveling over snow-covered ground. In winter they lived in low, half-buried homes with stone floors and massive whale bones to hold up the roof.

They were called the Thule Culture and were a highly successful people. Like the Dorset Culture they spread across North America to Greenland. Because they were skilled hunters at sea as well as on land, they were able to make full use of the limited Arctic resources.

Bone was used to make this comb and other finery by the Thule Culture around 1100 A.D.

The Names of Arctic People

Today the people of the Arctic are divided into many groups, or tribes. Each group has its own customs, identity, and name. They are known by the collective name "Inuit." This means "The People" and is the name most groups prefer to be called. A single person is an Inuk. The Inuit are sometimes called Eskimos. This word comes from native North Americans and means "eaters of raw flesh."

Living in the Arctic

Inuit are divided into many different tribes with names such as Saami, Chukchi, and Yup'ik. Between Alaska and Greenland there are about 78,000 Inuit. In Russia there are 600,000 to 700,000.

Surviving the Cold

In the Arctic there is no wood to build homes or light fires and no wool to make clothes. To survive, the Inuit had to find ways to keep warm.

Inuit dressed in the fur and skins of the animals around them, such as the Arctic foxes, caribou, polar bears, and seals. These kept them warm and dry. The warmest trousers were made from polar bear skins. Women often wore *kamiks*, which are long boots that reach the top of the leg. Kamiks were usually made from seal skin, which is waterproof.

Warm Homes

Some Inuit make snow houses. We call them igloos but to an Inuk an igloo is any type of house. Igloos are made by cutting blocks of snow and piling them up into a domed shelter. The entrance is a small tunnel, with a snow block for a door. Over the tunnel a block of transparent ice is set in the wall as a window.

In the 1500s Inuit women dressed like this. Her clothes are made from the skins of various animals. They are loose but very warm. The boots are made from sealskin. The hood of the parka is large enough to carry her baby, who is wrapped in moss for extra warmth.

Igloos are quick to build and often made when Inuit travel in winter. They are warmer than tents and much stronger against winter storms.

These homes could be so warm that people stripped off almost all their clothes when inside. Lamps gave light and heat for cooking, but most food was eaten raw or even frozen. Fresh meat and **blubber** from whales and seals was popular but in summer the diet was varied with moss, herbs, and birds' eggs.

In winter the Inuit traveled inland to hunt. This was when they lived in igloos. In summer they generally lived in tents made from caribou skin. Some Inuit made low houses with walls made from stone and whale bone. Skins were stretched over the walls as a roof. Several families would live together, using stone lamps that burned oil made from whale blubber.

Traveling on Ice

With no iron and little wood, Inuit used seal and whale bones tied with sinews to make sleds and boats.

A team of six to ten huskies was used to pull sleds across snow and sea ice. Huskies are strong and they can travel day after day even in winter. But in spring the snow melts and the sea ice breaks up. Inuit then used boats to hunt.

Inuit had two types of boat. The kayak was made by stretching seal skins over a frame of driftwood or bone. It was so light that it could be carried by one man. In summer, when the whales moved north into the Arctic Ocean to feed, they were hunted. Then a much larger boat, the umiak, was needed. It was nearly always rowed by women but was accompanied by men in kayaks who chased the seals or whales.

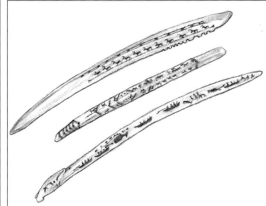

Ivory and Stone

There are no trees in the tundra to supply wood for building homes or making tools. No metals were mined, although some were obtained by trading with people living in the forests farther south. Instead the Inuit used what was available. There were plenty of bones from seals and whales. Ivory is much harder and came from tusks of walrus and some whales.

Ivory was used for making strong knives like those above, which would be used for building igloos. Large bones would be used for building homes and making sleds. Ivory, bone, and some types of stone were carved into the shapes of seals and birds, which the Inuit hunted or saw on their travels.

This 150-year-old kayak was made in Greenland. Great skill was needed when hunting seals among sea ice in a kayak.

Early Explorers: The Vikings

For centuries the only people in the Arctic were Inuit. Sometimes the climate became warmer. Then the Inuit, following their food supply, settled farther north. When it cooled again, they were forced south to find food. People of the Thule Culture were able to spread during just one such period. During another period of warm weather, which lasted several hundred years, people from Europe were able to spread to the Arctic.

The Vikings

The Vikings were good sailors. They lived in Scandinavia and built strong ships that could carry many goods.

They spread across northern Europe and reached Iceland in 860 A.D. When they arrived they found a small group of Irish monks. Soon more Vikings came and in less than a hundred years they had formed the Icelandic Althing, or parliament.

Vikings were the first Europeans to settle in the Arctic. They reached Iceland and Greenland in strong ships like this.

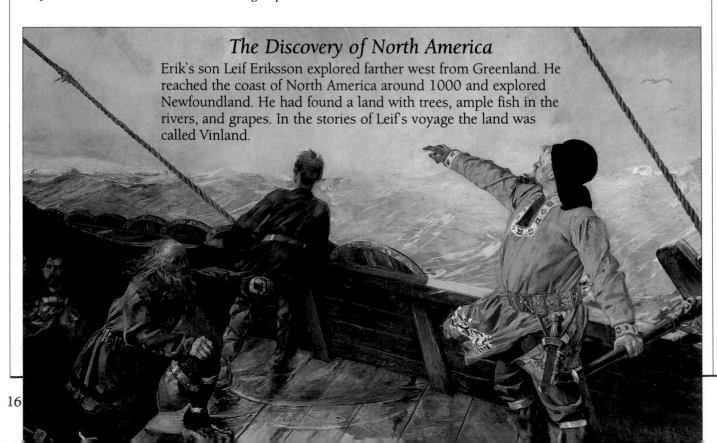

The Discovery of North America

Erik's son Leif Eriksson explored farther west from Greenland. He reached the coast of North America around 1000 and explored Newfoundland. He had found a land with trees, ample fish in the rivers, and grapes. In the stories of Leif's voyage the land was called Vinland.

These low stone walls built about 986 are all that remain of the Viking houses in Greenland. They were built by settlers with Erik the Red.

A Viking named Erik the Red was banished from Iceland in 982 for killing a man. The Viking stories told of lands to the west, so he decided to sail out to see if he could find them. After sailing for many days Erik sighted the snowy mountains of a new land.

Greenland

Erik's banishment was for three years. During the summers he explored the coast of this new land. In the winters he built a settlement. After the three years had passed, he returned to Iceland. He wanted to encourage more people to come to his settlement, called Eriksfjord. So Erik renamed the land Greenland to make it sound attractive.

Twenty-five ships set out for Greenland in 986, but only 14 arrived. The others were lost at sea. The **settlers** built stone houses with turf roofs near the seashore. They farmed cattle and sheep. They also caught fish and hunted seals and walrus. These activities gave them goods to trade with Europeans for grain. They met and traded with Inuit who had inhabited Greenland for thousands of years. But the Vikings never learned how to live off the land like the Inuit. Instead, each year a ship came from Norway, bringing them food they could not grow in Greenland.

Starved to Death

The Vikings came to Greenland during a period of warm winters and hot summers. About 1300 the weather became cooler and the winters longer. The settlement in Greenland had a terrible time. Calves born in winter were sickly and fewer seals were seen in summer. The supply ship from Norway came less and less often because more ice filled the seas. Finally the Vikings in Greenland all died.

The efforts of the Vikings in Greenland and North America were soon forgotten. It was another 300 years before Europeans "rediscovered" these lands.

The Traveling Abbot

Stories tell of an abbot, St. Brendan from Ireland, who went to Iceland about 500. With 17 monks he sailed in a large open boat made like a wicker basket and covered in oxhide. They carried wine and food for their journey, which took them to several islands on the edge of the Arctic. Tim Severin, a modern explorer, copied this voyage of St. Brendan in 1976 to see if it was possible. It was!

3 Arctic Exploration

The 14th to 16th Centuries

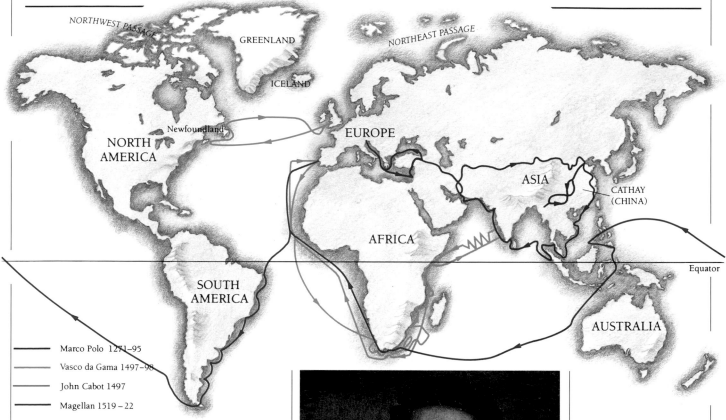

NORTHWEST PASSAGE

GREENLAND

NORTHEAST PASSAGE

ICELAND

Newfoundland

EUROPE

NORTH
AMERICA

ASIA

CATHAY
(CHINA)

AFRICA

SOUTH
AMERICA

Equator

AUSTRALIA

Marco Polo 1271–95
Vasco da Gama 1497–98
John Cabot 1497
Magellan 1519–22

Greenland was forgotten by the rest of the world after the Viking settlement died out. Stories about the Arctic became more unreal. From time to time a ship would be blown off course. The sailors would come home with descriptions of fierce tides and wild maelstroms (whirlpools). Most fearsome of all were stories of floating islands of ice that filled the seas. Fog often covered the sea, **mirages** twisted the horizon, and ships were smashed by ice.

Many were frightened by these stories and would not sail north. But the rich fishing grounds between Britain and Iceland were a great attraction to fishermen. In bad storms fishing boats sheltered in the bays of Iceland. The fishermen met the people of Iceland. By 1400 merchants from Bristol, England, heard of these people. They sent ships to trade with the Icelanders. But they came only in summer. Any ship caught by winter would be lost in storms or crushed by the ice.

Explorers searched for sea routes to Cathay (China). Portuguese and Spanish ships found routes around South Africa and South America, leaving the British and Dutch to look for an Arctic route.

Vasco da Gama (c. 1469–1525) was the first European to reach India by sea. He sailed around South Africa and proved there was no connection to Antarctica (see page 29).

Whaling was very dangerous work. The men in the small boat are trying to harpoon the whale to kill it.

Routes to Cathay

Most of Europe, however, was more interested in the Far East. Marco Polo had made an overland journey to Cathay (China) at the end of the 13th century. When he returned home to Venice he was a wealthy man. He told of silks and spices and many other goods for trade. The overland route was very dangerous. So Portuguese, British, and Dutch **navigators** searched for a sea route to Cathay.

Vasco da Gama (see left) sailed around South Africa in 1497–1498. He discovered a route for the Portuguese. Ferdinand Magellan's voyage (see page 28) in 1519–1522 was the first circumnavigation of the world. His route took him around South America. These voyages gave Portugal and Spain southern sea routes to Cathay. The French, Dutch, and English had to go north to find another way.

John Cabot (c.1425–1499) was a great explorer who lived in Venice. When he moved to England, he was sent in 1497 by King Henry VII to search for a route to Cathay. He sailed west and, 52 days after leaving England, he found land. In fact, he had rediscovered Newfoundland and was the first European to see it since the Vikings. But he had not found Cathay, and the English ignored his discovery.

The Arctic Route to Cathay

Many attempts were made during the next 300 years to find other ways to Cathay. The English and Dutch tried ways around North America in one direction, through the Northwest Passage. They also tried to find a way north of Russia in the other direction, the Northeast Passage. In the end neither way led them to the wealth of the Far East.

John Cabot, shown here with his son Sebastian, left Bristol in 1497 looking for an Arctic route to Cathay. He rediscovered North America.

The Search for the Northeast and Northwest Passages

The English and Dutch wanted sea routes to Cathay. They searched for a way through the Arctic. Hudson and Barents sailed north of Russia trying to find the Northeast Passage. About the same time Martin Frobisher, John Davis (c. 1550–1605), and later Hudson sailed around North America, looking for the Northwest Passage.

——— Muscovy Company Voyage 1553
——— Frobisher's Route 1576
——— John Davis' Route 1585
——— Barent's Route 1596
·········· Hudson's 'Hopewell' Voyage 1607
– – – Hudson's 'Hopewell' Voyage 1608
——— Hudson's Route 1610

Elizabeth I (1533–1603) wanted her sailors to find a sea route to Cathay. The overland route could not compete with the Portuguese and Spanish. The English thought there must be a route north of Russia. The Muscovy Company was formed to try and find it.

The Muscovy Company sent three ships in 1553. They soon met with tragedy. The crews of two ships froze to death north of Lapland. One ship struggled on to the White Sea. In terrible winter conditions the crew left the ship and traveled overland to Moscow. They did not find a way to Cathay, but they did establish a fur trade route with Russia.

Henry Hudson

Hudson was a remarkable English sea captain. He was born about 1550 and was trained as a navigator by the Muscovy Company. He was about 57 years old when the Muscovy Company sent him to look for a route to Cathay over the North Pole. Hudson left in 1607 in a tiny ship called *Hopewell* to explore the east coast of Greenland and the seas north of Spitzbergen. He saw that the seas there were full of whales. When the Dutch and English heard of this they started more than 200 years of whaling around Spitzbergen.

Barents was caught by ice near Novaya Zemlya.

Willem Barents

The Dutch also tried to find the Northeast Passage. In 1596 they sent Willem Barents. He thought he could sail around the ice-blocked sea by going farther north. But he was caught in the ice and his ship was crushed. Barents and his men were forced to camp for the winter on Novaya Zemlya. Barents died of cold and disease, but he is remembered as the first European to winter so far north of the Arctic Circle.

The Northwest Passage

During the 1570s and 1580s the English had also been sending ships to look for a Northwest Passage. Captain Martin Frobisher (c. 1535–1594) left in June 1576 and found a large strait west of Greenland. He thought this was the passage. Later it was shown to be a large bay in Baffin Island.

Ten years later another English seaman, named John Davis, set out in search of the passage. In the course of three voyages he sailed to what is now called Davis Strait and Baffin Bay.

By 1610 Hudson was working for the English again. He sailed on April 17 to explore the strait found by Davis. In June he found Hudson Strait and by autumn had crossed Hudson Bay to James Bay. There he spent the winter.

In the spring his crew **mutinied** because they thought they would die of starvation. They put Hudson, his son, and loyal crewmen in a small boat to fend for themselves. They were never seen again. The mutineers themselves were attacked by Inuit and only eight returned to London. On the voyage home they had to eat candles, grass, and bird skins to stay alive.

*F*robisher was one of several explorers who searched for the Northwest Passage. Later he fought against the Spanish Armada.

But the sailors were not hanged as was usual for mutineers. They were the only ones who knew the way to Hudson Bay. More voyages followed and showed that there was no way out of Hudson Bay to the west. During one of the trips, Lancaster Sound, north of Baffin Island, was discovered. Nobody realized it at the time but this was eventually shown to be the Northwest Passage.

The Hudson's Bay Company

The English and Dutch realized that no safe passage was to be found to Cathay by the Arctic. Instead, trade with America became more important to Europe. Hudson's discoveries had shown a way into Canada for traders. The Hudson's Bay Company was formed in 1670 and still trades today.

*H*udson (c. 1550–1611) was cast adrift by mutineers in 1611 after wintering on his ship in the Arctic. He and his son died near the large bay now named after him.

Siberia and Alaska

Vitus Bering was a Danish seaman who spent much of his life working for the Imperial Russian Navy. He made some of the greatest Arctic journeys of the 18th century, from one coast of Russia to the other, covering thousands of miles.

He was sent by the czar of Russia, Peter the Great, to explore the far eastern coast of Siberia. Russian fur trappers had spread to the east during the 1600s but no explorer had ventured so far. Peter the Great wanted to know whether Asia and North America were connected.

Bering started his first voyage from Moscow in 1725. First he **trekked** 5,000 miles across Siberia to the Pacific Coast. He set up a base at Kamchatka. Here he built ships to explore Eastern Siberia and the Arctic. In 1728 he sailed through the strait between Siberia and North America up to the Arctic Circle. During summer there is often fog over polar seas. In heavy fog, Bering missed the coast of North America. Thinking his work was done he returned to Moscow.

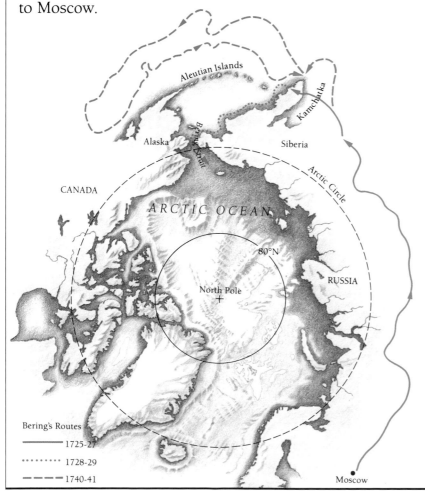

The Fur Trade

All around the Arctic are huge forests. These forests are rich in animals that are hunted by fur trappers. For most of the 1700s these forests supplied furs for clothing to wealthy Europeans. The sea and land routes discovered by Hudson and Bering provided a way of bringing the furs to Europe. Fur trappers spread far and wide across North America and Russia but unfortunately little is known of their journeys.

Wooden ships gave little protection to seamen sailing the icy Arctic Ocean.

Arctic Sailors

A sailor's life in the Arctic was a hard and often short one. Wooden ships were easily crushed by the ice. That is why most ships would only go to the Arctic in summer. The sailors fed on salt beef or pork, codfish, and dried peas. They also had some bread, cheese, and butter, and drank beer or water.

Scoresby (1798–1857) first went to the Arctic on his father's whaling ship. He explored many miles of Arctic coastline.

Unconvinced

The czar still believed that Siberia and North America were connected and wanted Bering to look again. Bering was delayed for six years in Russia but started again in 1740. He had to repeat his trek across Siberia and build new ships. But this time luck was with him and he explored the west coast of Alaska and the Aleutian Islands. He also mapped long stretches of the north Siberian coastline.

Disaster struck in 1741. On the return voyage from North America, their ship ran aground and was wrecked. Bering died of **scurvy**, bringing the total who perished on the expedition to 30. But Bering had discovered rich sealing and whaling grounds that brought wealth to the czar.

Bering was born in Denmark in 1681. He died of scurvy in 1741 on an island off the Alaska coast.

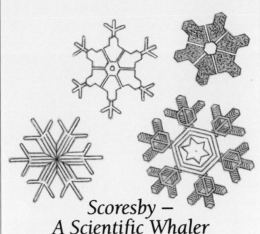

Scoresby – A Scientific Whaler

William Scoresby was one of the most successful Arctic whaling captains. He first sailed in ships owned by his father. He was well-educated and his books about the Arctic contained more scientific knowledge than any before. During his voyages he drew snowflakes and measured the temperature of the sea. He discovered that warmer water lies underneath the cold surface of the Arctic waters and he was the first to record the currents in the Arctic Ocean.

Sir John Franklin

Franklin made many Arctic expeditions. He died looking for the Northwest Passage.

Franklin's Routes

——————— 1818

– – – – – 1845

——————— Amundsen's Route 1903–6

In 1818 the British Navy sent two separate expeditions in four ships to the Arctic. War between England and France had just ended and the Navy had many spare ships. The admiralty used them to search again for the Northwest Passage.

John Franklin commanded one of these ships. He was born in 1786 and had joined the Navy at 14. He fought bravely at Copenhagen and Trafalgar and had been promoted to lieutenant by the time he left for the Arctic. For his first expedition Franklin went to the Greenland Sea.

Between 1819 and 1827 Franklin led overland expeditions to explore the Arctic coast of North America. He traveled more than 5,500 miles by canoe on grueling journeys that taught him much about Arctic travel. When he returned to England in 1827 Franklin was knighted for his discoveries.

Several overland expeditions crossed the Canadian Arctic to try to find the western end of the Northwest Passage.

Franklin's Last Journey

Franklin returned to the Arctic in 1845. He was 58 years old and he was again seeking the Northwest Passage. He took two strong ships named *Erebus* and *Terror*, with food for a three-year expedition and 134 officers and crew.

Franklin followed routes discovered in 1818 and 1819. The first winter was passed on Beechy Island, where three men died. Continuing the next summer (1846) he reached Victoria Strait but in the autumn the sea froze around his ships. The winter passed, but the ice did not melt the following summer and the ships stayed trapped.

Franklin died during that summer of 1847. Food ran low and 21 other men also died probably from hunger, scurvy and exhaustion. Still the ships were frozen in, despite all the efforts of the crew to free them. The surviving men were desperate. They decided to set out on foot across the sea ice to the mainland. They hoped to reach a trading post farther south, but they never arrived.

The Search for Franklin

Franklin had not returned to England by 1848 and a great search was started. During the next 10 years, 6 overland and 34 ship expeditions looked for Franklin. Only small clues were ever found. Some Inuit said they had seen a group of "white men" years before. A document was found on King William Island telling of Franklin's death. One Inuk had a silver plate marked with Franklin's name.

Island after island was searched looking for Franklin. During the search more of the Canadian Arctic was explored than ever before. Finally, one Northwest Passage was found by Captain Robert McClure. Another was discovered by John Rae, a Scottish explorer who also found evidence of Franklin's failed expedition. But it was not until 1903–1906 that the Norwegian explorer Roald Amundsen finally sailed through a Northwest Passage.

Sir William Parry

Parry made five expeditions to the Arctic, including an attempt to reach the North Pole. Born in 1790, he joined the Navy, like Franklin. With Sir John Ross in 1818 he discovered the first part of the Northwest Passage. On later expeditions he successfully helped to locate the **North Magnetic Pole**.

Lady Jane Franklin

Franklin's second wife, Jane, complained to the British admiralty that they were doing too little to search for her husband. She used her own money to advertise the search and insisted that the government set a reward for information about her husband. The first information concerning Franklin was learned from the Inuit by Dr. John Rae in 1854. Eventually, the admiralty lost interest in tracing her husband so she organized an expedition (1857–1859) led by Sir Francis McClintock. He finally confirmed Franklin's death.

Peary and Cook: The Race for the North Pole

After Franklin died a new type of explorer came to the Arctic, whose only aim was to be the first to reach the North Pole. Two Americans, Robert Peary and Frederick Cook, both claimed to be the first. Neither claim has been proved beyond doubt.

Learning to Live Like an Inuk

Peary made eight Arctic expeditions. He trained as a **surveyor** before joining the U.S. Navy as an **engineer**. He was 34 when he went on his first Arctic expedition in 1891. On it Peary showed that Greenland was an island by mapping its northern coast. He confirmed that the way to the Pole must be over the sea ice of the Arctic Ocean.

Peary worked closely with the Inuit. He organized whole villages to help him. He learned Inuit ways of sledding with dogs. He also learned how to live off seals, polar bears and caribou as the Inuit did.

Off to the North Pole

Peary left New York in July 1908. He sailed in the *Roosevelt* to his winter camp on the north coast of Ellesmere Island. From here the North Pole lay across 480 miles of sea ice. The *Roosevelt* was frozen into the ice and gave the expedition a home for the winter.

Peary left on February 22, 1909, for the North Pole. He took 6 Americans, 17 Inuit, 19 sleds, and 133 dogs. Every few days a group of men and sleds was sent back to the *Roosevelt*. When Peary was 155 miles from the Pole, he sent back the last group.

Peary claimed to be the first person at the North Pole. He is wearing the clothes he wore on the polar expedition.

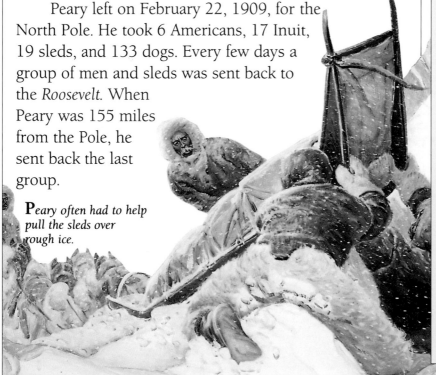

Peary often had to help pull the sleds over rough ice.

Other Attempts at the North Pole

In 1827 Parry sailed and sledded to 82° 45′N but he was still 500 miles from the Pole. Sir George Nares led the British Arctic Expedition of 1875–1876, which set a new record of 83° 20′N.

Six years later this was broken by Adolphus Greely, an American, who reached 83° 24′N.

A Norwegian named Fridtjof Nansen made an unusual expedition. He built a special ship, *Fram*, which could resist crushing by sea ice, and set out in 1893. Nansen let his ship get frozen into the ice north of Siberia. *Fram* drifted across the Arctic Ocean for three years and broke free just north of Spitzbergen. During the drift, Nansen attempted to sled to the Pole, reaching 86° 14′N in 1895.

Peary went on with his companion, Matthew Henson, four Inuit, and the best dogs.

Peary reached the North Pole on April 6, 1909, and camped there for the night. The next day he took photographs of the group. Turning for home, they reached the *Roosevelt* in 16 days.

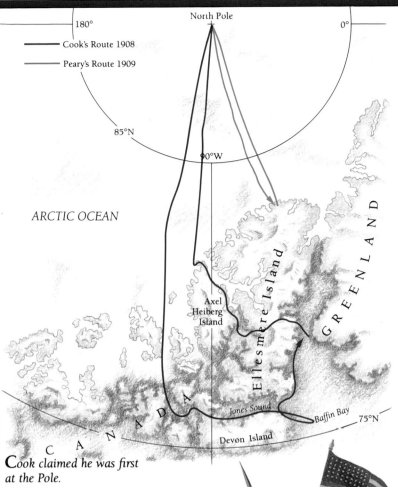

*C*ook claimed he was first at the Pole.

Cook Reaches the Pole First?

Just five days before news of Peary's success reached New York, Cook announced that he had reached the North Pole a year before. In 1908 Cook led a hunting expedition to Greenland. During the trip he sent a message home that he was going to the North Pole.

Cook had learned, like Peary, to travel with Inuit. He left Greenland on February 18, 1908, with 10 Inuit, 11 sleds, and 105 dogs. He crossed Ellesmere Island and sledded over the sea ice to the tip of Axel Heiberg Island. From there Cook sledded over sea ice toward the North Pole. Cook and two Inuit with 26 dogs arrived at the North Pole on April 21, 1908.

While Cook was returning over the sea ice, the current swept him off course. He missed food supplies left on Axel Heiberg Island for his return. Instead he landed farther to the west. He continued on the sea ice to Jones Sound, where he camped for the winter. By now all the dogs had died or been shot. Cook and his companions started back across Ellesmere Island to Greenland in the spring of 1909. They arrived half starving and in a terrible state.

Why Peary and Cook Are Doubted

People question whether Peary could have traveled as fast as he said he did across the sea ice. Neither he nor Cook could produce convincing navigational records. Cook's diaries of the trip were left in safekeeping in Greenland and were never seen again. The two stories are not easy ones to verify. However, most experts now believe Peary was the first to make it.

4 Antarctic Exploration

Defining the Limits

The discovery of Antarctica is a very different story from that of the Arctic. Arctic history starts 20,000 years ago. Antarctica was not even seen until 1820.

The Antarctic Ocean surrounds Antarctica. Early humans and many animals could not cross it. This is why Antarctica remained isolated.

Antarctica is the only continent Europeans truly "discovered." People were living in most countries when the Europeans arrived. However, there were no people living in Antarctica when the first explorers arrived there.

The Unknown Land

The Greeks thought that there must be a large land in the south to balance the lands they knew in the north. So the idea of a vast southern continent came about. The land at the South Pole was called the "opposite of the Arctic," or Antarctic.

Many of these ideas were forgotten during the **Middle Ages**. Only the Arabs kept the Greek teachings alive, but few Europeans understood Arabic. Latin was the language of teachers and the churches in Europe. It was not until the 15th century that the Greeks' ideas were translated into Latin. Nothing was known about the southern continent of the Greek teachings. It was called the "unknown land in the south," which is *Terra Australis Incognita* in Latin.

The Greeks thought Terra Australis *extended from the equator to the South Pole. The map above shows the Greek ideas.*

The Portuguese explorer Magellan showed that a strait separates South America from Terra Australis.

How map makers saw Antarctica in 1570.

Exploring the Antarctic Ocean

People imagined that this land stretched from the equator to the South Pole. Maps from the 15th century show it linked to Africa, Asia, and South America. Africa was the first to be separated. Vasco da Gama sailed around the south coast of Africa in 1497 looking for a way to India. In 1519 Magellan found a way around South America into the Pacific. Magellan had seen land to the south of his route. The map makers then showed *Terra Australis Incognita* covering much of the Pacific Ocean and only separated from South America by a strait.

In 1577 Captain Francis Drake left England officially to discover *Terra Australis*, but also to attack Spanish ships and lands. He sailed south of Magellan's route around Cape Horn. He showed that South America is separated from the southern lands by what he called Drake Passage. He then sailed across the Pacific Ocean and marked the land on maps. Drake had proved the map makers wrong again.

A modern Polynesian sailing canoe

A Frozen Sea

None of these explorers saw Antarctica. The only hint of a frozen land near the South Pole came from an early Pacific Ocean legend. The people of Polynesia, who live on islands in the Pacific Ocean, tell a story of a great leader. This leader, Ui-te-Rangiora, sailed his canoe as far south as a frozen ocean in 650.

Near Misses

Captain Cook sailed around Antarctica but never saw the mainland. He made many other discoveries, including South Georgia. In 1779 he was killed in Hawaii while returning from a voyage to the Arctic.

In January 1773 Cook's ships (below) were the first to cross the Antarctic Circle. They weathered furious and dangerous Antarctic storms, and sailed seas full of ice during their voyage.

A century and a half after Drake, two French explorers, Jean-Baptiste Bouvet and Yves-Joseph de Kerguelen-Tremarec, on separate voyages thought they saw Antarctica. Most people were only interested in trade and took little notice of their discoveries. Map makers still drew *Terra Australis Incognita* covering most of the Antarctic Ocean.

A Scientific Sailor

James Cook was born in England in 1728. He went to sea when he was 18 years old and worked on ships transporting coal. Cook was very smart and an excellent seaman. He was also a good scientist. Cook later joined the British Navy, which sent him on several expeditions. On one he charted New Zealand and Australia's east coast.

Cook left England on July 13, 1772, with two ships, *Resolution* and *Adventure*, bound for Antarctica. He took 27 tons of crackers and thousands of pieces of salted pork. Even so, his crew ate meat only four days a week.

Cook was told to sail "as near to the South Pole as possible." He was also told to claim any land he discovered in the name of the King of England. During the next three years Cook sailed 60,140 miles and went around the continent of Antarctica.

Farthest South

Cook sailed as far south as he could, often along the edge of the **pack ice** surrounding Antarctica. Ice would form on the ropes and sails, making the ships likely to **capsize** and impossible to handle. The seas were full of icebergs and pack ice. One serious collision could sink a ship. On January 17, 1773, Cook's ships were the first to cross the Antarctic Circle.

Cook went to New Zealand for the Antarctic winter. When the weather became better he sailed south again. He sailed along the ice edge but did not see any land. He did show that if there was a continent it was much smaller than the land drawn on maps. He also showed that the places seen by Bouvet and Kerguelen-Tremarec were only islands and not the edge of Antarctica.

The Second Circumnavigation

Forty-five years passed before Antarctica was sailed around again. Many **sealers** went to islands such as South Georgia looking for fur seals. They probably saw many parts of Antarctica, but there are few records of their voyages.

It took another clever seaman to attempt what Cook had done. He was a Russian naval officer named Captain Thaddeus Bellingshausen. He left Russia in 1819 with two ships, *Mirny* and *Vostok*.

Bellingshausen knew of Cook's voyage. Where Cook was forced north by ice or poor weather, Bellingshausen would try to go south. For days all he could see was pack ice. On January 20, 1820, *Mirny* crashed into a large **floe**. The crew survived and pressed on. Several months later in Australia, they discovered that the ice had made a 3-feet-long hole in the ship's side. Only a layer of **tarred canvas** had kept the sea from rushing in.

Bellingshausen discovered several new islands. On January 27, 1820, he came within 20 miles of the coast of Antarctica. At that point the coast is made up of long, low ice cliffs. To Bellingshausen they looked just like a line of icebergs. Because he did not see any rocks, he did not recognize it as the continent.

Cook and Bellingshausen had similar bad luck. Both made extraordinary voyages around Antarctica. Somehow they survived seas full of ice and Antarctic storms. Yet neither could say that they had seen the continent.

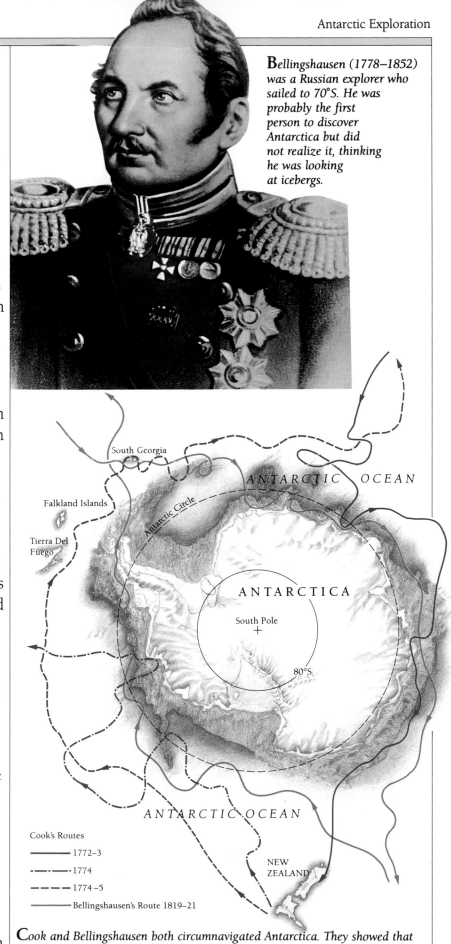

Bellingshausen (1778–1852) was a Russian explorer who sailed to 70°S. He was probably the first person to discover Antarctica but did not realize it, thinking he was looking at icebergs.

Cook's Routes

——————— 1772–3

—·—·—·— 1774

— — — — 1774–5

——————— Bellingshausen's Route 1819–21

Cook and Bellingshausen both circumnavigated Antarctica. They showed that the continent was far smaller than previously thought. Both discovered several new islands, many full of fur seals.

The First Antarctic Winter

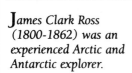

James Clark Ross (1800-1862) was an experienced Arctic and Antarctic explorer.

Borchgrevink (1864-1934) wintered here on Cape Adare. He was the first to do so.

Between 1778 and 1815 most of Europe was at war.

Exploration of Antarctica was left to sealers, whalers, and private expeditions.

Searching for Seals

Cook's voyage had alerted sealers to the rich harvest in the Antarctic. Sealers came from Britain and the United States and later from Australia and France. Some are remembered in the names of places they found. James Weddell reached the southern point 74° 15′S, a record at the time, in the Weddell Sea. John Biscoe sailed around Antarctica, finding Enderby Land and the Biscoe Islands. He was the first person to confirm seeing the mainland of East Antarctica, describing black mountains showing through the ice.

One sealing company had a keen interest in scientific exploration. The Enderby brothers told their captains to explore new lands while seeking seals. For 40 years their ships led Antarctic exploration.

Magnetism

It was scientists in Europe working on **magnetism** who led the next stage of exploration. Britain, the United States, and France each sent expeditions to find the South Magnetic Pole.

James Clark Ross was an experienced Scottish explorer who had first gone to the Arctic when he was 12. In 1831 he located the North Magnetic Pole. In 1839 the British admiralty sent him to the Antarctic to seek the South Magnetic Pole. His ships, *Erebus* and *Terror*, were well built. They were strong enough for Ross in the Antarctic, but were later lost during Franklin's expedition to the Arctic (see pages 24–25).

Ross's ships were the first to break through the pack ice that lies between New Zealand and Antarctica. He discovered the Ross Sea but was stopped by a vast, 100–150 foot high ice barrier, the Ross Ice Shelf. This was a most important discovery. It gave a possible way to reach the South Pole.

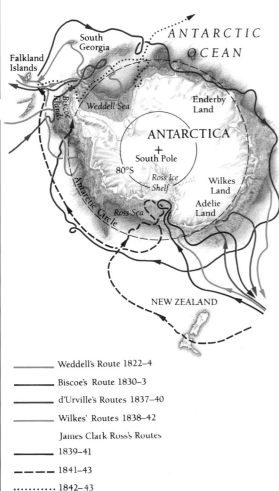

South Georgia
Falkland Islands
Biscoe Islands
Weddell Sea
ANTARCTIC OCEAN
Enderby Land
ANTARCTICA
+ South Pole
80°S
Ross Ice Shelf
Wilkes Land
Adélie Land
Antarctic Circle
Ross Sea
NEW ZEALAND

———— Weddell's Route 1822–4
———— Biscoe's Route 1830–3
———— d'Urville's Routes 1837–40
———— Wilkes' Routes 1838–42
James Clark Ross's Routes
———— 1839–41
– – – – 1841–43
·········· 1842–43

Ross discovered Ross Island (above) where Scott and Shackleton later started their attempts to reach the South Pole.

Wilkes and d'Urville

Ross did not find the South Magnetic Pole, but he did sail around Antarctica. Two other expeditions left about the same time to seek the South Magnetic Pole. Charles Wilkes left the United States with five ships and Jules Dumont d'Urville left France with two ships. Both failed. Yet these voyages did complete much more of the Antarctic map. Wilkes Land, Adélie Land, and the Adélie penguin were major discoveries. Dumont d'Urville named the last two after his wife, Adélie.

The First Antarctic Winter

European explorers first wintered in the Arctic in 1819. It was another 80 years before explorers stayed for the winter in the Antarctic. Carsten Borchgrevink, a Norwegian, raised funds in Britain for a private expedition to winter in Antarctica.

On his first expedition he found a suitable spot at Cape Adare, Victoria Land. When he returned in February 1899 it took ten days to unload the specially built huts he brought with him. Twice during the winter the huts were nearly lost. Once a bedside candle started a fire and another time fierce winds threatened to blow the huts down.

Ten people stayed for the winter. They used dog sleds for the first time in Antarctica and made many short trips. Later, back on board their ship, they visited the Ross Ice Shelf. Borchgrevink showed that a well-equipped expedition could stay in Antarctica all year. He proved that a person could travel over Antarctic ice. This set the scene for the next stage of Antarctic discovery—the race for the South Pole.

Hunting Whales

When Ross came back to England he said the Ross Sea was full of whales. Fifty years later this information was used to start a whaling industry. By the 1900s thousands of whales were being killed each year around the Antarctic. So many were hunted that they almost became extinct.

The Siege of the South Pole

The British wanted to be the first at the South Pole. They sent an expedition commanded by a young naval lieutenant named Robert Falcon Scott.

Scott attempted to sled to the Pole in 1902. The first stage was across the Ross Ice Shelf, known then as the Great Ice Barrier. Then he had to cross a mountain chain to reach the inland ice plateau. No one knew what lay beyond.

Scott, with Ernest Shackleton and Edward Wilson (see page 37), sledded to 82° 16'S. Their food ran low and Shackleton was so ill from scurvy that Scott feared for his life. They turned back.

The Plateau

Shackleton returned in 1908. He took just 29 days to pass the earlier record by using ponies to pull the sleds. Climbing a mountain a week later, he looked down on the largest glacier ever seen. It flowed from the inland plateau down into the Barrier and it was the way to the Pole.

On the plateau, Shackleton had to turn back on January 7, 1909, just 110 miles from the Pole. They had been eating half-rations for days and could go no farther.

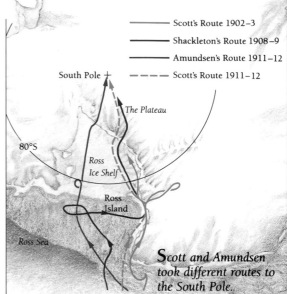

—— Scott's Route 1902–3
—— Shackleton's Route 1908–9
—— Amundsen's Route 1911–12
- - - Scott's Route 1911–12

South Pole

The Plateau

80°S

Ross Ice Shelf

Ross Island

Ross Sea

Scott and Amundsen took different routes to the South Pole.

Huts were built to withstand Antarctic weather. During winter, explorers' equipment was prepared.

Scott took Siberian ponies for his expedition. They needed exercising during the winter.

Scott Tries Again

Scott made a second attempt in 1910. He built a hut on Ross Island and during winter prepared food and equipment for the trip. Scott took ponies, motor sleds, and some dogs to help pull the sleds.

He set out on November 1, 1911, following Shackleton's route from 1908–1909. At the plateau, Scott and four men went on. The others returned to the hut to wait.

Warm clothing is needed for the Antarctic plateau. Here are examples of some of the things Scott took.

The Cold Plateau

These five men had pulled their sleds since leaving the barrier. They had a terrible time on the plateau. Temperatures below −4°F and strong winds delayed them. They were exhausted by their long journey.

Twenty miles from the Pole they saw a black flag. Soon they realized that the Norwegians had beaten them to it.

The Final Blizzard

Scott and his men were stunned, and in very low spirits turned for home. Short of food and fuel, they suffered frostbite and scurvy. The weather worsened. A blizzard on the barrier in March 1912 delayed them just 11 miles from supplies. Scott and the last of his companions died in their tent during the blizzard.

The Norwegian Success

The Norwegian expedition leader, Roald Amundsen, was an experienced polar traveler. He had wintered on board ship in the Antarctic in 1898 and made many journeys in Norway's mountains. He had learned that dog sledding was the most efficient way to travel over snow.

In 1910 Amundsen planned to go to the North Pole. Then he heard that Peary and Cook had reached it. He decided in secret to attempt the South Pole. Only after he had set out did he send a message to Scott: "Beg leave inform you proceeding Antarctica. Amundsen."

A New Route

Amundsen built a hut on the Ross Ice Shelf well to the west of Ross Island. He was already 60 miles closer to the Pole than Scott, but he had to find a new route.

Dogs can travel in colder temperatures than ponies. When spring came, Amundsen was able to set out 11 days before Scott. Amundsen had few problems on his journey. Despite taking a new route he made a fast time. The Norwegians reached the South Pole on December 14, 1911, a month before Scott. Amundsen was a tough and smart explorer who won Antarctica's greatest prize.

Amundsen (1872–1928) used Arctic methods of travel for his attempt at the South Pole. He explored a new route and reached the Pole a full month before Scott. He disappeared in the Arctic on another expedition.

A Forgotten Continent

Expeditions had reached the South and Magnetic Poles but vast areas were still unexplored. British and Norwegian whaling companies plundered Antarctic waters. Meanwhile a series of small expeditions came to Antarctica. Many were privately organized and many used aircraft for the first time to help in exploration.

Aircraft opened up Antarctica. In 1928 Richard Byrd led an American expedition to the Ross Ice Shelf and made the first flight to the Pole in 1929. Byrd with Floyd Bennett was also the first to fly over the North Pole in 1926. Five years later Byrd returned with a much larger expedition. He extended his old base called Little America. It now housed more than over 40 men in a dozen buildings. Byrd mapped the edge of the Ross Ice Shelf.

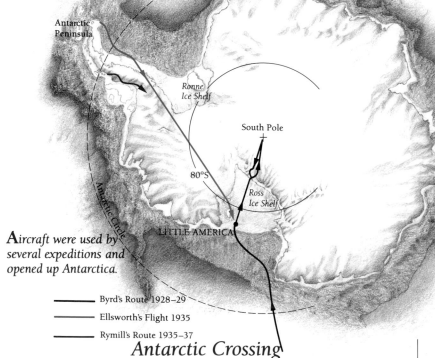

Aircraft were used by several expeditions and opened up Antarctica.

——— Byrd's Route 1928–29
——— Ellsworth's Flight 1935
——— Rymill's Route 1935–37

Shackleton's Bravery

Ernest Shackleton organized a third expedition. In 1914–1916 he attempted to cross Antarctica. But his ship, *Endurance*, never landed. It was crushed by ice in the Weddell Sea. Only Shackleton's skill as a leader saved him and his crew. By sleds and boats he reached Elephant Island. Shackleton and five others sailed 800 miles to South Georgia to arrange a rescue. Shackleton brought all his men home safely.

Antarctic Crossing

From the other side of Antarctica a small expedition prepared a plane for a historic flight. In November 1935 Lincoln Ellsworth and a pilot set out to fly across Antarctica. He believed it would take 14 hours but blizzards forced them to land and wait for good weather. Twenty-two days later they ran out of fuel 16 miles from Little America. Ellsworth and his pilot had to walk there. Byrd had left two years before. The two men had to wait a month for their ship to sail around Antarctica to collect them.

Dogs and Planes

John Rymill was an Arctic explorer. During one trip he planned to explore the Antarctic Peninsula. Dog sledding and aircraft were now the usual way to travel. Rymill used both to prove that the peninsula was not an island as had once been thought. It was part of Antarctica.

After World War II the United States had a large navy and in 1946 mounted the largest Antarctic expedition ever seen. The expedition was called Operation Highjump. Richard Byrd commanded 13 ships, 25 aircraft, and more than 4,000 men. Little America was reopened for the summer but this time nobody stayed for the winter. Aircraft were used to map and photograph more than half of the Antarctic ice cap. Much of it had never been seen before.

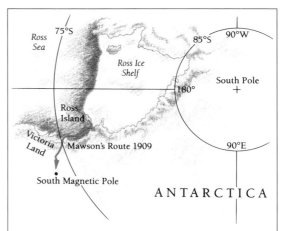

Rymill took an expedition to the Antarctic Peninsula. He was a thorough and methodical leader.

Whiling away winter in the hut. Using dogs, Rymill's expedition sledded over 1,240 miles.

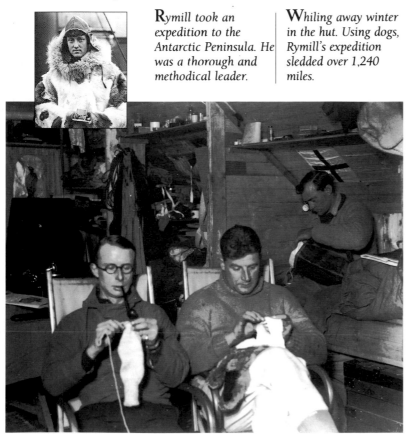

International Expeditions

Operation Highjump was followed by Operation Windmill. This was a much smaller expedition. Both expeditions marked the start of a new type of Antarctic exploration. Antarctica is so big that only governments could afford the large expeditions needed. But even governments had trouble finding the money and for some expeditions several countries had to work together.

Mawson reached the South Magnetic Pole.

Mawson and the South Magnetic Pole

When Shackleton set out for the South Pole he sent another group, including the Australian Douglas Mawson, to seek the South Magnetic Pole. They left the hut in late September and found the magnetic pole on January 16, 1909, after a 1,000-mile march. Mawson returned to Antarctica in 1911 commanding an Australian expedition, which returned to the magnetic pole and discovered many new areas.

Mawson returned to Antarctica in 1929. Now Sir Douglas Mawson, he led a joint British, Australasian, and New Zealand expedition that set the style for all future scientific work and confirmed that the Antarctic really was a continent.

International Cooperation

A small expedition in 1949 changed Antarctic exploration. Three countries pooled their resources and scientists. This Norwegian-British-Swedish expedition made many scientific and geographical discoveries in Dronning Maud Land.

We refer to most expeditions by the names of their leaders. From Cook onward we remember the name of the person in command. This expedition is remembered as the start of international cooperation in Antarctica.

Claiming Antarctica

When a new area of Antarctica was discovered, it was claimed in the name of the expedition's country. Ross claimed Victoria Land for Britain. The United States claimed all the land seen during Operation Highjump; explorers even dropped markers by aircraft as part of their claim. By World War II several countries, including Argentina, Chile, Britain, France, and Norway, had claimed parts of Antarctica.

Scientific Exploration

Scientists changed all that. In the early 1950s a group of European and American scientists planned the International Geophysical Year, or IGY. They wanted to measure the activity of the sun all around the world. It was also decided to make exploration of Antarctica part of the IGY.

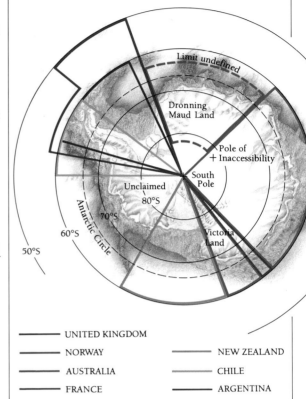

—— UNITED KINGDOM	
—— NORWAY	—— NEW ZEALAND
—— AUSTRALIA	—— CHILE
—— FRANCE	—— ARGENTINA

Today, seven nations claim parts of Antarctica. They have agreed not to dispute these claims while the Antarctic Treaty is in force.

Crossing Antarctica

The first crossing was made by Sir Vivian Fuchs. On an expedition from 1955 to 1958 he used dogs and tractors called Snowcats to cross Antarctica in 99 days. Several other expeditions have now crossed Antarctica. Some have repeated the journeys of Scott and Amundsen. Although modern equipment was used, these expeditions still were dangerous.

In 1992 Sir Ranulph Fiennes and Dr. Michael Stroud attempted another crossing. They successfully pulled their own food across Antarctica without the help of motor vehicles, ponies, or dogs.

Fuchs at the Pole during the first Antarctic crossing

Unloading supplies at a scientific station. Almost a ton of food is needed for each person for a year.

Scientific stations are run all year in Antarctica. Their studies have shown how important Antarctica is to our well-being.

The IGY ran from mid-1957 to December 1958. Twelve countries opened stations in Antarctica or on nearby islands. The United States chose the South Pole. An entire station was brought in by aircraft in 1956. It took 84 flights to fly in all the materials and equipment. Nobody had stood at the South Pole since Scott and Amundsen sledded there. The next person to arrive came 44 years later by airplane.

The Soviets chose the Pole of Inaccessibility for one of their stations. This is the hardest point in Antarctica to reach. They used tractor trains to bring in the materials for their station. The tractors took several weeks to travel the 900 miles from the coast. Dozens of tractors went in a line together, sometimes more than a mile long. They called their station Vostok. The coldest temperature in the world, − 128°F, was later recorded there.

The Antarctic Treaty

The IGY was a great success. Nations worked together as never before. Then, in 1961 an agreement called the Antarctic **Treaty** was reached. Scientists could explore Antarctica without worrying about land claims. Nobody minded who found or saw what. This ideal situation has continued for over 30 years.

As the needs of scientists have changed, stations have opened or been abandoned. In 1993 there were 48 scientific stations in the Antarctic. These stations were run by 20 different nations. Some other nations worked only during the Antarctic summer. They ran another 20 stations.

5 Polar Politics

In the Arctic

Several countries have shorelines on the Arctic Ocean. They are some of the major industrial nations of the north such as Russia, the United States, and Canada. At the beginning of this century much of the Arctic was unclaimed. There seemed little of value there. Throughout the 18th and 19th centuries Alaska provided a fur trade for Russia. Russia ruled Alaska but when the furs ran out, Russia sold the land to the United States.

By 1950 the Arctic countries had sliced up the Arctic like a cake. Each nation claimed a piece that stretched from its coastline to the North Pole. The USSR and North America faced each other across the Pole.

Polar Rule from Afar!

The largest Arctic island is ruled from far away. Greenland has been connected with Denmark since Viking times. Through all the changes in Europe and Scandinavia, Denmark ruled Greenland. Greenland has moved toward independence from Denmark since World War II. In 1979 home rule was given to Greenland. Fishing now gives Greenlanders their main income. It will be some time before they can cut all **economic** ties with Denmark.

Five nations border the Arctic Ocean. Some argue over the position of their borders in the Arctic Ocean. At stake are large areas of continental shelf rich in fish and probably containing oil and gas.

An International Land

One group of islands in the Arctic is unusual. Whalers worked from Spitzbergen for many centuries. There was still no clear owner as late as 1900. When English and American companies began coal mining, things came to a head. A treaty was signed in 1920 that gave Spitzbergen to Norway. The Spitzbergen Treaty allowed all those who signed it to visit and mine on Spitzbergen. Spitzbergen is a little like an international territory. The only other place run on similar lines is the Antarctic.

Native Rights

The Arctic peoples have suffered in the same way as native people elsewhere. Eurasians and North Americans have often come to explore or hunt wildlife. For hundreds of years they killed whales and seals and hunted animals for fur. During this century they have mined for oil and minerals.

The Inuit of Alaska, Canada, and Greenland have formed a group. By coming together they have had more effect on the governments that rule their land. The Alaskan Inuit made an agreement with the U.S. government in 1971. It gave them a billion dollars and about a tenth of the land.

The Arctic is rich in minerals, coal, oil, and gas. Much damage has been done to the tundra extracting this wealth, using mines like the one above.

The Inuit have suffered at the hands of Eurasians and North Americans for hundreds of years. Only now are payments being made for the use of their land.

This was in exchange for the right of the government to mine. Other groups are trying to make similar agreements.

Arctic people in the former USSR were treated much worse. Things have improved but the new government of Russia is too poor to help much. There is little chance the Siberian natives will receive any money for the **minerals** taken from their lands.

Arctic Science

Science has a long history in the Arctic. Some of the earliest expeditions had scientific aims. Finding the North Magnetic Pole was one such aim.

In 1990 the International Arctic Science Committee, called IASC for short, was formed. Thirty years earlier, a similar group had formed in the Antarctic. The IASC was based on the older Antarctic group. Scientists in Antarctica had inspired the Antarctic Treaty. It is just possible that scientists in the Arctic could have a similar effect.

In the Antarctic

The politics of Antarctica could not be more different from the Arctic. No countries border Antarctica as they do the Arctic.

How to Claim a New Land

Various countries claim parts of Antarctica. Some areas, such as the Antarctic Peninsula, are claimed by more than one country. Different ways can be used to prove a piece of land belongs to a country. One way is to say that country was the first to discover it, which is how most explorers claimed the land.

Another way is to say how much land the country has seen. In 1960 the United States claimed that its explorers had flown over more of Antarctica than any other nation. Dropping markers from aircraft was also used. Wherever the markers fell, land was claimed.

The Antarctic Treaty has brought peace to Antarctica. At the South Pole, station flags of treaty members fly as a reminder of the benefits of cooperation.

Slices of the Cake

The first formal claim to part of Antarctica was made by the British government in 1908. In all, seven nations have made claims, some of which overlap. In 1950 much of Antarctica was still unknown, yet anger was building among various countries with overlapping claims. The United States and former USSR were also thinking of making claims. The United States wanted to claim all of Antarctica.

The Scientists' Success

Fortunately, scientists took the lead and worked together for the International Geophysical Year (see page 38). The scientists agreed they had no interest in claiming land.

The year was so successful that the scientists wanted to continue. They formed the Scientific Committee for Antarctic Research, SCAR for short, in 1958. This group has kept nations working together for Antarctic science.

The scientists' success led directly to the Antarctic Treaty. Twelve countries worked very hard for two years to prepare the treaty. It was signed on December 1, 1959, and came into force in 1961. The treaty is one of the great international agreements.

Antarctica still has the feeling of a remote and wild place. Many people think it should be kept that way and made into a world park.

Polar Tourism

Many people want to visit the polar regions. The scenery and wildlife in the Arctic and Antarctic are so special that people pay thousands of dollars to see them. If they are not to damage what they have come to see, tourists need to be controlled. The Antarctic Treaty Environmental Protocol will help to do this. In the Arctic each country has different laws and control is less easy.

Mining and Tourists

An issue not covered by the treaty was mining. In the 1980s an attempt was made to write a mining treaty for Antarctica. Although it had many good parts, in the end it was not passed. In the 1970s and 1980s tourists began to come to Antarctica more often. About the same time people began to think that the treaty was going to expire in 1991. In fact the treaty will remain in effect forever.

Nongovernment groups became involved in Antarctica for the first time. Greenpeace suggested that Antarctica should be made a world park. This would allow scientific research to continue. It would also allow a small amount of tourism. Mining would be forbidden.

Protection of Antarctic Wilderness

By 1990 nearly 40 countries had signed the Antarctic Treaty. They felt that the world park idea would have no advantage over the existing treaty. Instead they agreed to strengthen the treaty. In 1992 they agreed to a new section called the Environmental **Protocol**. This made much stricter rules about what could be done in Antarctica. Part of the protocol was a ban on mining for 50 years.

Footprints in the Snow

Four great rivers flow from Russia into the Arctic Ocean. They carry waste and pollution from factories far to the south. The Arctic is surrounded on all sides by nations with factories—the United States, Canada, countries of Europe, and Russia. The health of the Arctic Ocean is very important to people who live in these areas.

Early explorers such as Amundsen and Franklin generally only left a few footprints in polar snows. Scientists have now discovered how much we are affecting the poles. Our lifestyles have even changed the composition of the atmosphere over the poles. This is called the ozone layer. A "hole" was first noticed in the Antarctic. We may even be causing the whole earth to become warmer. Such warming could melt the ice caps.

The poles are important to the state of the earth. They have been called a **life-support system** of the planet. They are used to check the effects we are having. A 6,500-foot-long ice core has been drilled at Vostok in Antarctica. At its deepest point the ice dates back to 160,000 years ago. Traces of factory-made chemicals and pollution have been found in the core. These are from things never used in Antarctica. They were used elsewhere in the world and brought to Antarctica by winds.

The Arctic is also being affected. In summer a haze lies over the Arctic Ocean, caused by dust and pollution from factories around the Arctic. The native Arctic peoples have been disturbed many times by Europeans, first by explorers claiming the land and pushing them out. More recently, mining has completely changed the Inuit lifestyle. Animals and plants have been disturbed just as much.

People have been careless in the polar regions, often leaving trash behind. Thankfully, this is now changing.

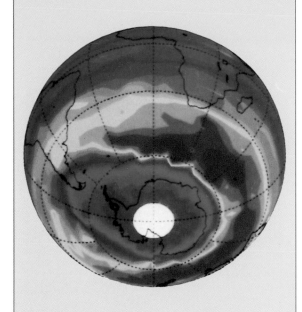

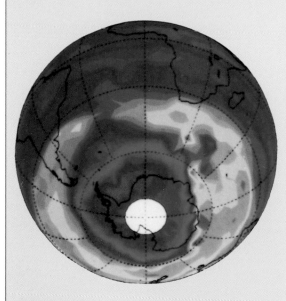

Satellite pictures show the ozone hole over Antarctica in 1991 and 1992. The highest levels of ozone are shown in red and the lowest levels in pale blue.

The Ozone Hole

Ozone protects us from harmful rays in sunlight. Some of the chemicals we use are destroying the layer of ozone. We notice it because we burn and our skin is damaged more easily by the sun now. Animals and plants cannot use sunscreen like we do. They are also damaged by the harmful rays in sunlight. Each of us can help by only using things that do not harm the ozone layer.

What to Do?

Scientists have shown just how important the poles are to the health of the earth. People need to be aware of the far north and south. The discovery of the ozone hole has helped to start this awareness.

Waste and pollution from our factories reach the poles by water and by air. We must look carefully at everything we use. We just do not know what may end up at the poles.

Arctic oil extraction brought changes to the lifestyles of people and animals. Great care is needed when building pipelines like this.

Visitors

An easier problem to tackle is the effect of people visiting the poles. The Antarctic Treaty includes some control of visitors. The Arctic does not have such a treaty, and laws vary among Arctic nations. They have little control over the Arctic Ocean.

Scientists have not always given the best lead in the Arctic and Antarctic. They have often made a mess around the bases where they work. Not all of their science has been good for the animals.

Scientists and visitors need to respect the Arctic and Antarctic environments. The Arctic and Antarctic are largely untouched compared with developed countries. We should make sure that no more than a few footprints are left in the snow when we leave.

Arctic	Antarctic	Other
c.2500 B.C. Dorset Culture develops in Alaska.		c.2500 B.C. Pythagoras believes the earth is a globe.
c.500 A.D. St. Brendan sails to Iceland.	c.150 A.D. Greek maps show Antarctica stretching from equator to South Pole.	c.230 A.D. The Great Wall of China is built.
860 The Vikings reach Iceland.	c.650 Ui-te-Rangiora sails south to the frozen sea.	622 Muhammad founds the religion of Islam in Arabia.
986 Erik the Red lands on Greenland with settlers.	1000 Greek teachings about Antarctica are kept alive by Arab scholars.	1066 The Normans conquer England.
c.1400 English merchants trade with Iceland.	1497 Vasco da Gama sails around the south coast of Africa showing that it is not connected to Antarctica.	1492 Christopher Columbus reaches the Americas.
1553 Muscovy Company sends three ships to find the Northeast Passage.	1519-22 Ferdinand Magellan proves South America is separated from Antarctica.	1530s The European slave trade begins across the Atlantic Ocean.
1607 Henry Hudson leaves on his first Arctic voyage.	1577 Francis Drake leaves England to seek Terra Australis.	1569 Gerardus Mercator publishes his first world map.
1725 Vitus Bering starts on his first trans-Russia expedition.	1739 Jean-Baptiste Bouvet thinks he has discovered Antarctica.	1707 Union of Scotland and England.
1740 Bering's second expedition.	1772-1775 James Cook circumnavigates Antarctica.	1775 American Revolution begins.
1819 John Franklin leads his first Arctic expedition.	1820s Thaddeus Bellingshausen sights Antarctic continent during circumnavigation.	1789 French Revolution begins.
1831 James Clark Ross discovers the North Magnetic Pole.	1841 Ross discovers the Ross Ice Shelf – the route to the South Pole.	1861 American Civil War begins.
1847 Franklin dies searching for the Northwest Passage.	1899 Carsten Borchgrevink leads the first expedition through winter on Antarctica.	1895 Guglielmo Marconi invents the radio.
1895 Fridtjof Nansen sleds to 86°N during Fram voyage.	1902 Robert Scott reaches 82°S on his first attempt at the South Pole.	1903 Wright brothers' first flight.
1908 April, Frederick Cook claims to have reached the North Pole.	1909 Ernest Shackleton reaches 88°S on an attempt at the South Pole.	1905 Start of the Russian Revolution.
1909 April, Robert Peary claims to have reached the North Pole.	1911 December, Roald Amundsen reaches the South Pole.	1911 Chinese Revolution begins.
1920 Spitzbergen Treaty is signed.	1912 March, Robert Scott and companions die during return from South Pole.	1914-1918 World War I.
1926 Richard Byrd claims to have flown over the North Pole.	1929 November, Richard Byrd claims to have flown over the South Pole.	1930 Gandhi starts campaign of civil disobedience in India.
1948 Soviet airplane lands on sea ice at the North Pole.	1935 November, Lincoln Ellsworth flies across Antarctica.	1939-1945 World War II.
c.1950 Arctic countries claim slices of the Arctic Ocean.	1956 United States builds a scientific station at the South Pole.	1957-1958 International Geophysical Year.
1959 U.S. nuclear-powered submarine surfaces at the North Pole.	1955-1958 Sir Vivian Fuchs crosses Antarctica.	1957 USSR launches the first satellite, Sputnik I.
1969 Wally Herbert makes first crossing of the Arctic Ocean.	1961 Antarctic Treaty is signed.	1961 Berlin Wall is built in Germany.
1971 Alaskan Inuit Land Claims settlement.	1970s Tourists start to visit Antarctica.	1968 Dr. Martin Luther King Jr. is assassinated.
1979 Greenland is granted home rule by Denmark.	1979 Air New Zealand aircraft carrying 257 tourists crashes on Ross Island.	1975 U.S. withdraws from Vietnam.
1990 International Arctic Science Committee is formed.	1992 Mining is banned in Antarctica for 50 years.	1991 Berlin Wall is destroyed, and East and West Germany are reunited.

Glossary

B

blubber: the thick layer of fat underneath the skin of seals, whales, and penguins. Animals living in the polar regions need this fat to keep warm. It also provides a food reserve in winter.

C

capsize: to overturn a boat in the water. Kayaks are so light they easily capsize.
climate: the weather of a country or continent.

E

economic: everything concerned with money.
engineer: a person who learns to make and repair machines.

F

floe: a small piece of **sea ice** floating on the sea.

H

hemispheres: two divisions into which the globe is divided; north and south. The equator is the dividing line.

I

ice cap: a large and permanent area of ice that covers the land. Very large ice caps, such as the one that covers Antarctica, are often called ice sheets. The ice in an ice cap flows like the ice in a glacier but at a much slower rate.

L

life-support system: a phrase used to describe something that is very important in keeping people of the earth healthy.

M

magnetism: a force that cannot be touched or seen but which attracts certain metals. It causes the magnet in a compass to point to the **North Magnetic Pole**.
Middle Ages: the period of four hundred years around 1100 to 1400 A.D. It was a time in Europe when civilization appeared to stand still.
minerals: valuable substances mined from the ground. Coal, iron, and many metals are minerals.

minute: a measurement that equals one mile in geography. Sixty minutes is one degree. Latitude and longitude are divided into degrees (60 miles) and minutes (1 mile).
mirage: a trick of light that makes things seem nearer and larger than they are. Most common in deserts and the polar regions.
mutiny: revolt by sailors or soldiers against their officers. Orders are ignored and the officers are often chased away.

N

navigators: sailors who can safely bring a ship across the ocean, even through uncharted waters.
Neanderthal: an ancient type of people who lived over 100,000 years ago. They were probably forced into extinction about 40,000 years ago when modern people developed.
North Magnetic Pole: one of two areas of the earth that acts like a magnet. The poles of the earth magnet, just like an iron one, have a north and a south end but they do not line up with the North and South Poles. The North Magnetic Pole lies just north of Canada and is where the needle of a compass points.

O

ocean current: the flow of cold and warm water through the sea like wind blowing across the land.

P

pack ice: a type of **sea ice** that has broken into large **floes**. It is blown together by winds and swept along by currents and can trap and crush small ships.
polar regions: the frozen lands and cold seas around the North and South Poles.
protocol: the part of a **treaty** or agreement that gives the detail of how the agreement should be worked.

S

scurvy: a lack of vitamin C in the diet causing poor health which can eventually kill. Teeth and gums become rotten, arms and legs are painful. Scurvy can be cured by eating fresh fruit and vegetables. Many early sailors drank lime juice to try to prevent scurvy.

sea ice: ice which forms on the top of the sea in the **polar regions** when it is cold enough. It may become 6–10 feet thick and can be piled up by storms into 15-foot ridges.
sealer: somebody who hunts seals for a living. Seals are hunted for their fur and for oil that can be extracted from their **blubber**.
settlers: people who move to a new country looking for a fresh place to live.
surveyor: somebody who makes maps and charts of land and seas.

T

tarred canvas: canvas treated with pitch or tar to make it waterproof. It can be used in an emergency to stop a leak in a ship's hull.
treaty: an agreement made between governments from different countries to bring about peace or better relations.
trek: a long expedition across rough countryside made on foot.
tundra: the vast and often level lands that surround the Arctic Ocean. The climate is too cold for trees to grow and in summer the tundra is covered in pools and streams formed by the melting winter snow. Worst of all, the tundra is home to billions of midges and biting flies.

Index

Numbers in **bold** indicate an illustration. Words in **bold** are in the glossary on page 47.